BUILDING BLOCKS

Math Matters!

COUNTING

in Our World

By Naomi Osborne

Cavendish Square
New York

There are so many things to count!

You can count pieces of pizza. There is one piece of pizza.

1

You can count books.
There are two books.

1
2

You can count cupcakes.
There are three cupcakes.

1
2
3

You can count apples.
There are four apples.

2
1
3
4

You can count
birds in the sky.
There are five birds.

1
2
3
4
5

You can count dogs.
There are six dogs.

1
2
3
4
5
6

You can count eggs.
There are seven eggs.

1
2
3
4
5
6
7

You can count blocks.
There are eight blocks.

1
2
3
5
4
6
7
8

You can count pieces
of candy.
There are nine pieces
of candy.

1
2
3
4
5
6
7
8
9

Counting is fun!

Published in 2022 by Cavendish Square Publishing, LLC
243 5th Avenue, Suite 136, New York, NY 10016

First Edition

Website: cavendishsq.com

Cataloging-in-Publication Data

Names: Osborne, Naomi.
Title: Counting in our world / Naomi Osborne.
Description: New York : Cavendish Square, 2022. | Series: Math matters!
Identifiers: ISBN 9781502656490 (pbk.) | ISBN 9781502656513 (library bound) | ISBN 9781502656506 (6 pack) | ISBN 9781502656520 (ebook)
Subjects: LCSH: Counting–Juvenile literature. | Addition–Juvenile literature.
Classification: LCC QA113.O785 2022 | DDC 513.2'11–dc23

Editor: Vanessa Oswald
Copy Editor: Nathan Heidelberger
Designer: Deanna Paternostro

The photographs in this book are used by permission and through the courtesy of: Cover (top) Mike Price/Shutterstock.com; cover (bottom) Icatnews/Shutterstock.com; pp. 3, 23 Studio.G photography/Shutterstock.com; p. 5 New Africa/Shutterstock.com; p. 7 Klochkov SCS/Shutterstock.com; p. 9 Lithiumphoto/Shutterstock.com; p. 11 solomonphotos/Shutterstock.com; p. 13 rck_953/Shutterstock.com; p. 15 OlgaOvcharenko/Shutterstock.com; p. 17 Victoria ArtWK/Shutterstock.com; p. 19 goodmoments/Shutterstock.com; p. 21 Sofiia Tiuleneva/Shutterstock.com.

Printed in the United States of America